Simply
STARGAZING

Your Guide to the Stars, Moon, and Night Sky

Adventure Quick Guides

Simply Stargazing offers a wealth of information to help you find your way across the sky and identify what you see with confidence.

THIS HANDY QUICK GUIDE INCLUDES:

- Six full-sky star maps to help you identify stars and constellations
- Charts that introduce each season's most prominent constellations and how to navigate from one landmark to another
- Information about the signs of the zodiac and the ecliptic, the apparent path of the sun and planets across the sky
- A large map of the moon to help you identify its most prominent features
- Tips to find and identify the planets, bright deep-sky objects and the best times of the year to watch for "shooting stars"

WHAT TO BRING

Warm clothes: Night air can be chilly, even in the summer

A blanket and pillow: Most of your time will be spent looking up

This Quick Guide: Identify what you are seeing

A red light: Preserve your night vision

Binoculars: If you own a pair, bring them along. Binoculars help you spot dimmer stars, see rich detail in the Milky Way, and view craters on the moon

PRESERVE YOUR NIGHT VISION

Find a place sheltered from the glare of lights, and use a red light to view this Quick Guide. Our eyes take 20 minutes to adapt to the dark. White lights "reset" our night vision—soft red lights don't. Inexpensive red lights are easy to find, or make your own by applying red film from a craft store over a regular flashlight.

Jonathan Poppele is an award-winning author, naturalist, and educator who works to help people connect more deeply to themselves, others, and the natural world. He is an active member of the Minnesota Astronomical Society, founder of the Minnesota Wildlife Tracking Project, and Head Instructor of the Center for Mind-Body Oneness. Find him online at www.jonathanpoppele.com.

PHOTO CREDITS All the moon images in the guide were created by Ernie Wright from the Scientific Visualization Studio at NASA's Goddard Space Flight Center. **Mercury:** NASA/Johns Hopkins University Applied Physics Laboratory/Carnegie Institution of Washington **Mars**: NASA/JPL-Caltech/USGS **Jupiter:** NASA/JPL/Space Science Institute **Saturn:** NASA/JPL/Space Science Institute **Venus, Milky Way, Andromeda Galaxy, Orion Nebula, Praesepe and Pleiades:** Shutterstock

Cover and book design by Jonathan Norberg

Simply Stargazing: Your Guide to the Stars, Moon, and Night Sky
Copyright © 2018 by Jonathan Poppele
Published by Adventure Publications, an imprint of AdventureKEEN
310 Garfield Street South, Cambridge, Minnesota 55008
(800) 678-7006
www.adventurepublications.net
All rights reserved. Printed in China.
ISBN 978-1-59193-581-0 (pbk.)

USING THE STAR MAPS

The star maps in this Quick Guide show how the night sky appears every four hours, all year long. To choose the best map, find the current month on the chart below, scan over to the time when you're observing, then turn to the page for that map. For example, if you're observing in late April, the early spring map shows how the sky appears at 9:00 pm; the late spring map shows the sky at 1:00 am, and the summer map shows the sky at 5:00 am. The chart is blank when the sky is too bright for stargazing. The center of each map shows the sky directly overhead. The edge shows the horizon, marked with compass directions. Hold the book in front of you and turn it until the direction on the bottom matches the direction you are facing. The stars on the map will now line up with the stars in the sky. The star maps show the sky for 40° north latitude—a line through Pittsburgh, Denver, and Lebanon, KS (the geographic center of the contiguous US)—and are suitable for up to 15° above or below this line.

Month	Early Spring	Late Spring	Summer	Fall	Late Fall/ Early Winter	Late Winter
Early Jan	3:00 am				7:00 pm	11:00 pm
Late Jan	2:00 am	6:00 am				10:00 pm
Early Feb	1:00 am	5:00 am				9:00 pm
Late Feb	12:00 am	4:00 am				8:00 pm
Early Mar	11:00 pm	3:00 am				7:00 pm
Late Mar	11:00 pm*	3:00 am*				
Early Apr	10:00 pm*	2:00 am*				
Late Apr	9:00 pm*	1:00 am*	5:00 am*			
Early May		12:00 am*	4:00 am*			
Late May		11:00 pm*	3:00 am*			
Early Jun		10:00 pm*	2:00 am*			
Late Jun		9:00 pm*	1:00 am*			
Early Jul			12:00 am*	4:00 am*		
Late Jul			11:00 pm*	3:00 am*		
Early Aug			10:00 pm*	2:00 am*		
Late Aug			9:00 pm*	1:00 am*	5:00 am*	
Early Sep				12:00 am*	4:00 am*	
Late Sep				11:00 pm*	3:00 am*	
Early Oct				10:00 pm*	2:00 am*	6:00 am*
Late Oct				9:00 pm*	1:00 am*	5:00 am*
Early Nov				7:00 pm	11:00 pm	3:00 am
Late Nov				6:00 pm	10:00 pm	2:00 am
Early Dec	5:00 am				9:00 pm	1:00 am
Late Dec	4:00 am				8:00 pm	12:00 am

* = Daylight Savings Time

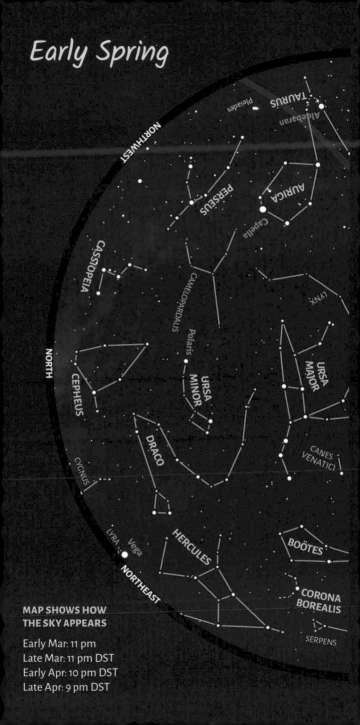

Early Spring

MAP SHOWS HOW
THE SKY APPEARS

Early Mar: 11 pm
Late Mar: 11 pm DST
Early Apr: 10 pm DST
Late Apr: 9 pm DST

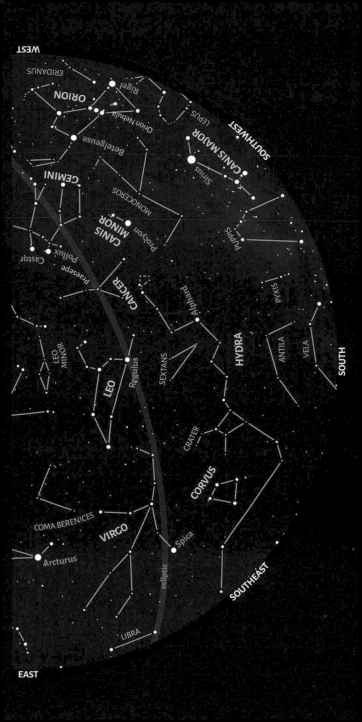

Late Spring

NORTHWEST

CANCER

Praesepe

Pollux

Castor

GEMINI

LEO MINOR

LYNX

Capella

AURIGA

URSA MAJOR

CAMELOPARDALIS

PERSEUS

Polaris

URSA MINOR

NORTH

CASSIOPEIA

DRACO

CEPHEUS

LACERTA

Vega

Deneb

LYRA

CYGNUS

NORTHEAST

VULPECULA

SAGITTA

DELPHINUS

Altair

**MAP SHOWS HOW
THE SKY APPEARS**

Early May: 12 am DST
Late May: 11 pm DST
Early Jun: 10 pm DST
Late Jun: 9 pm DST (during twilight)

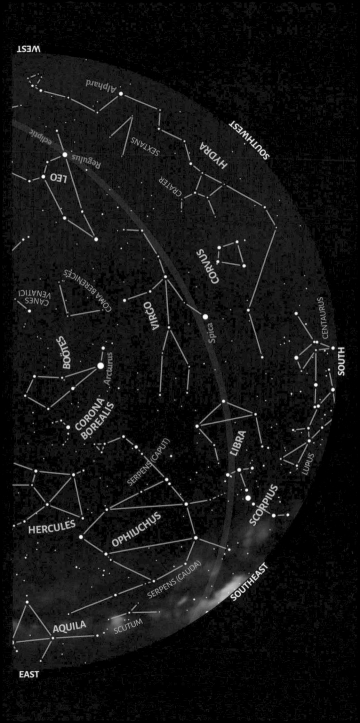

Summer

LEO

LEO MINOR

COMA BERNICES

CANES VENATICI

BOÖTES

NORTHWEST

URSA MAJOR

DRACO

LYNX

URSA MINOR

Polaris

CAMELOPARDALIS

NORTH

CEPHEUS

CASSIOPEIA

Deneb

CYGNUS

PERSEUS

LACERTA

ANDROMEDA

Andromeda Galaxy

TRIANGULUM

NORTHEAST

PEGASUS

PISCES

**MAP SHOWS HOW
THE SKY APPEARS**

Early Jul: 12 am DST
Late Jul: 11 pm DST
Early Aug: 10 pm DST
Late Aug: 9 pm DST

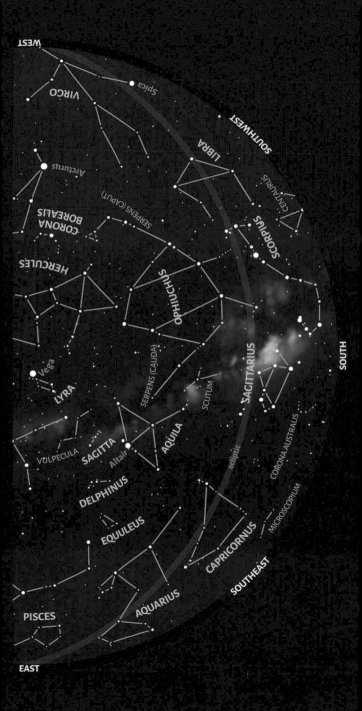

Fall

SERPENS (CAPUT)

BOÖTES

CORONA BOREALIS

NORTHWEST

HERCULES

LYRA
Vega

CYGNUS

Deneb

URSA MAJOR

DRACO

URSA MINOR

Polaris

CEPHEUS

LACERTA

NORTH

CASSIOPEIA

Andromeda Galaxy

CAMELOPARDALIS

ANDROMEDA

LYNX

PERSEUS

TRIANGULUM

Capella

AURIGA

NORTHEAST

Pleiades

**MAP SHOWS HOW
THE SKY APPEARS**

TAURUS

Aldebaran

Early Sep: 12 am DST
Late Sep: 11 pm DST
Early Oct: 10 pm DST
Late Oct: 9 pm DST
Early Nov: 7 pm
Late Nov: 6 pm

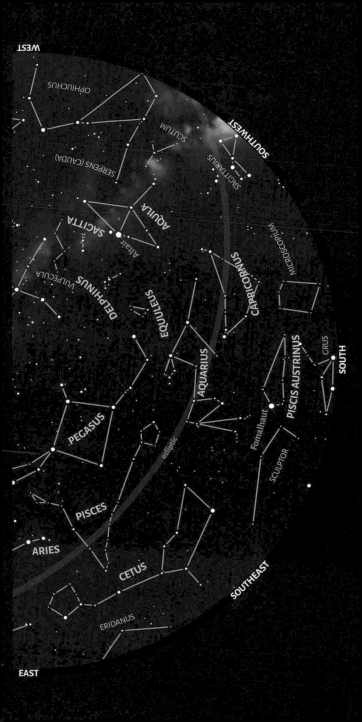

Late Fall /
Early Winter

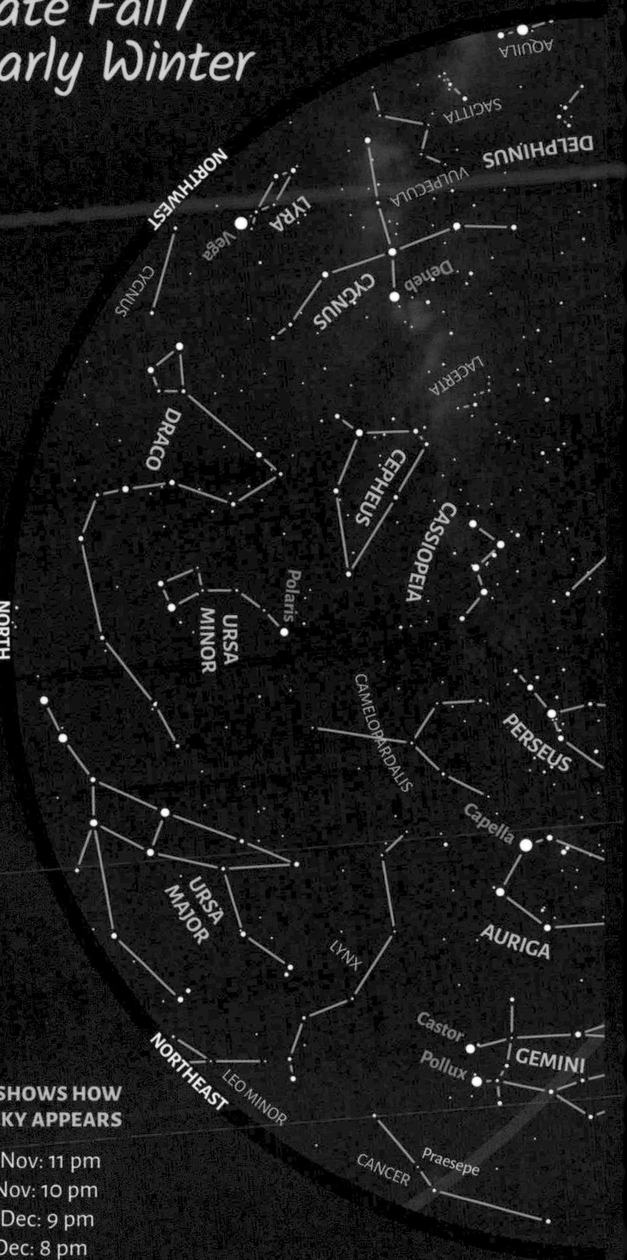

AQUILA

SAGITTA

DELPHINUS

VULPECULA

NORTHWEST

LYRA

Vega

CYGNUS

Deneb

CYGNUS

LACERTA

DRACO

CEPHEUS

CASSIOPEIA

Polaris

URSA
MINOR

NORTH

CAMELOPARDALIS

PERSEUS

Capella

URSA
MAJOR

LYNX

AURIGA

Castor

Pollux

GEMINI

NORTHEAST

LEO MINOR

MAP SHOWS HOW
THE SKY APPEARS

Early Nov: 11 pm
Late Nov: 10 pm
Early Dec: 9 pm
Late Dec: 8 pm
Early Jan: 7 pm

CANCER

Praesepe

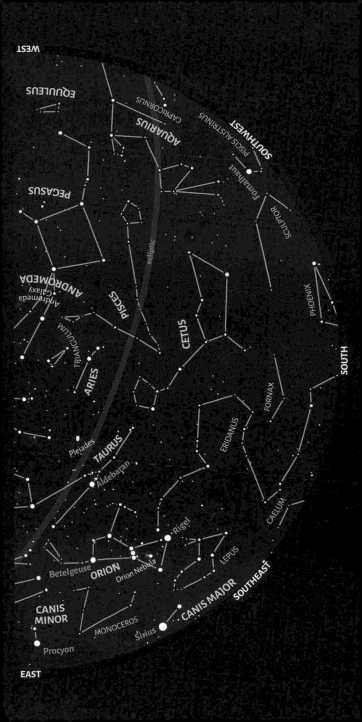

Late Winter

PEGASUS

ANDROMEDA

NORTHWEST

LACERTA

TRIANGULUM

Andromeda Galaxy

CASSIOPEIA

PERSEUS

CEPHEUS

CAMELOPARDALIS

Capella

NORTH

Polaris

DRACO

URSA MINOR

LYNX

URSA MAJOR

LEO MINOR

NORTHEAST

CANES VENATICI

COMA BERENICES

**MAP SHOWS HOW
THE SKY APPEARS**

Early Jan: 11 pm
Late Jan: 10 pm
Early Feb: 9 pm
Late Feb: 8 pm
Early Mar: 7 pm

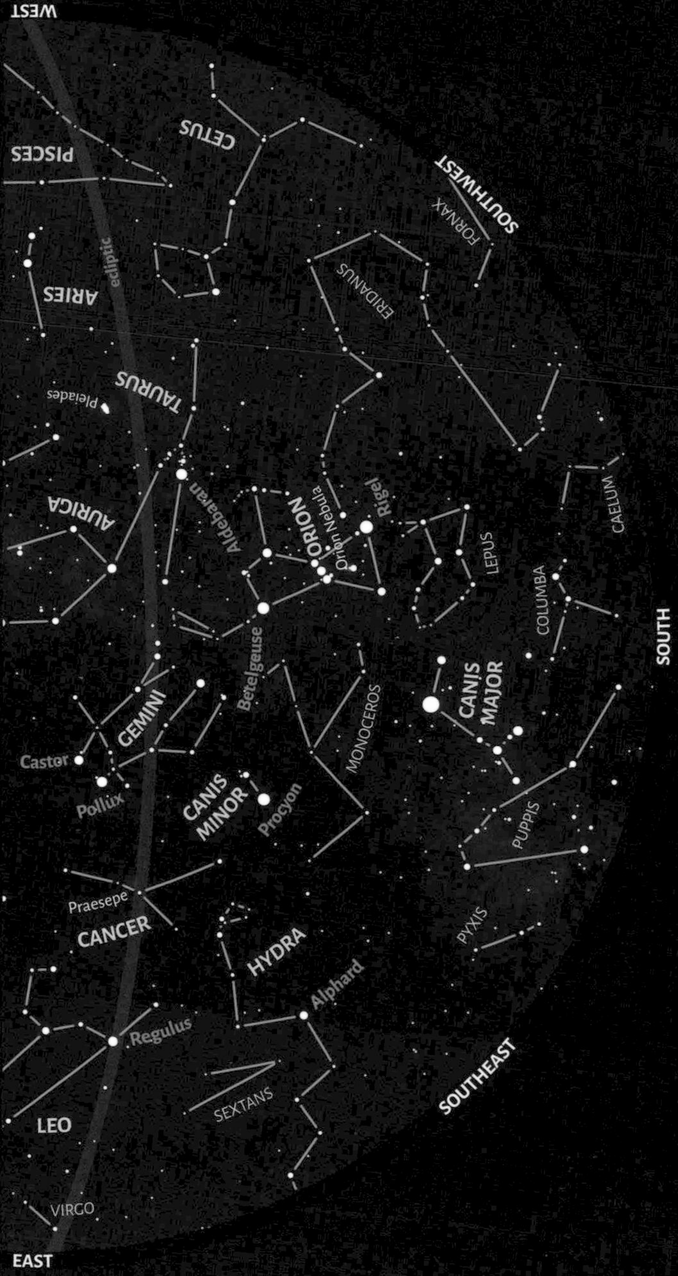

WEST
PISCES
CETUS
SOUTHWEST
FORNAX
ARIES
ecliptic
ERIDANUS
TAURUS
Pleiades
AURIGA
Aldebaran
ORION
Orion Nebula
Rigel
CAELUM
LEPUS
COLUMBA
SOUTH
Betelgeuse
MONOCEROS
CANIS MAJOR
GEMINI
Castor
Pollux
CANIS MINOR
Procyon
PUPPIS
Praesepe
CANCER
HYDRA
PYXIS
Alphard
Regulus
SOUTHEAST
SEXTANS
LEO
VIRGO
EAST

THE BIG DIPPER

The Big Dipper, part of Ursa Major, is the most famous star pattern in the north and will guide you across the spring sky.

1. To Find True North and More

· The front edge of the dipper's ladle points to Polaris, the North Star.
· Follow that line past Polaris to Cepheus.
· The wide "W" of Cassiopeia lies on the opposite side of Polaris from the handle of the dipper.

2. Arc to Arcturus, Speed on to Spica, Continue to Corvus

· Follow the arc of the dipper's handle to Arcturus in Boötes.
· Keep following the curve to Spica, the bright star in Virgo.
· Continuing this curve leads to Corvus near Hydra's tail.

3. Follow the Ladle

· The back edge of the ladle points south toward Regulus in Leo and beyond to Alphard in Hydra.
· The top of the ladle points to brilliant Capella in Auriga
· A diagonal line through the ladle points toward the twins of Gemini: Pollux and Castor.

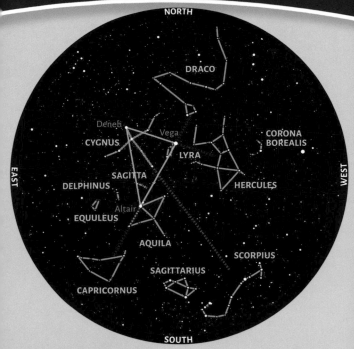

THE SUMMER TRIANGLE

A trio of bright stars from summer's three "bird" constellations form a large triangle that frames the Milky Way: Deneb, in Cygnus, the swan; Altair, in Aquila, the eagle; and Vega, meaning "vulture", in Lyra.

1. A Helpful Landmark

· Tiny Sagitta, the arrow, sits inside the Summer Triangle, just above Altair. To the southeast are miniature Delphinus, the dolphin, and Equuleus, the little horse.
· The top edge of the Triangle points west to Hercules.
· Beyond Hercules is the beautiful curve of Corona Borealis, the northern crown.
· The western edge of the Summer Triangle points north to the head of Draco, the dragon, and south to Capricornus.

2. The Northern Cross

· The brighter stars of Cygnus are known as the Northern Cross.
· The long line of the cross points south along the Milky Way toward Sagittarius and Scorpius.

THE GREAT SQUARE OF PEGASUS

Although not exceptionally bright, the Great Square frames a relatively empty region, making it the most prominent feature in the fall sky.

1. The Andromeda Group

The Great Square points to all the constellations in the legend of Princess Andromeda.

· The western edge points north to Cepheus, Andromeda's father.
· The eastern edge points north to Cassiopeia, her mother, and south to the tail of Cetus, the sea monster.
· The northeast corner marks the beginning of Andromeda herself, who in turn points to the hero, Perseus.

2. "A" for Andromeda

· Andromeda looks like a tall, narrow letter "A" turned on its side.
· Just northeast of the "A" is M31, the Andromeda Galaxy.
· The larger point in the "W" of Cassiopeia also points toward M31.

ORION, THE WINTER TRIANGLE, & THE GREAT HEXAGON

Orion, the hunter, is one of the most familiar patterns in the heavens. Let the famed hunter show you around the winter sky.

1. Orion's Belt---

· Orion's Belt points southeast to the brightest star in the sky: Sirius in Canis Major. The belt points northwest toward the famous Pleiades star cluster in Taurus.

2. Winter Triangle---

· Bright-orange Betelgeuse marks Orion's shoulder. Together with Sirius and Procyon, it forms a pattern known as the Winter Triangle, which lies along the faint "winter" Milky Way and frames the dim constellation Monoceros.

3. Great Hexagon---

· Betelgeuse is also the center of a huge star pattern known as the Great Hexagon. The corners of the Hexagon are marked by Capella in Auriga, Aldebaran in Taurus, Rigel in Orion, Sirius in Canis Major, Procyon in Canis Minor, and Pollux in Gemini.

The sun, moon, and planets are among the brightest objects in our sky. Since these bright objects move against the background stars, they do not appear on the star maps in this guide. Their movement, however, is always within a narrow band around the path of the sun. This path, called the "ecliptic," is shown on the star maps. If you see a bright "star" near the ecliptic that's not on the maps, it's probably a planet.

The Location of the Moon

The moon is large, bright, and instantly recognizable—so bright that it washes out dim stars, making it harder to recognize some constellations. Knowing the location of the moon in the sky makes it easier to identify nearby constellations. If you know the moon's phase, you can use this chart to find its position on the ecliptic. Many newspapers and calendars give the moon's phase. To begin, find the current date on the chart, which marks the position of the sun. The first quarter moon always appears three months ahead of this date; the full moon will be six months ahead or behind this date; and the third quarter moon is three months behind this date. For example, a full moon on April 1 will appear in Virgo.

Find Your True Zodiac Sign

Around 2,500 years ago the Greeks divided the band around the ecliptic into twelve equal "houses" and named each after a nearby constellation. They called this band the zodiac. Over time the Earth's axis has wobbled, causing the sun's position along the ecliptic to shift. The chart on the next page shows the current position of the sun throughout the year. You can use it to find your "True Sun Sign," the exact position of the sun on the day you were born, and compare it to the traditional dates shown below.

♈ Aries
Mar 21–Apr 19

♋ Cancer
Jun 21–Jul 22

♎ Libra
Sep 23–Oct 22

♑ Capricorn
Dec 22–Jan 19

♉ Taurus
Apr 20–May 20

♌ Leo
Jul 23–Aug 22

♏ Scorpio
Oct 23–Nov 21

♒ Aquarius
Jan 20–Feb 18

♊ Gemini
May 21–Jun 20

♍ Virgo
Aug 23–Sep 22

♐ Sagittarius
Nov 22–Dec 21

♓ Pisces
Feb 19–Mar 20

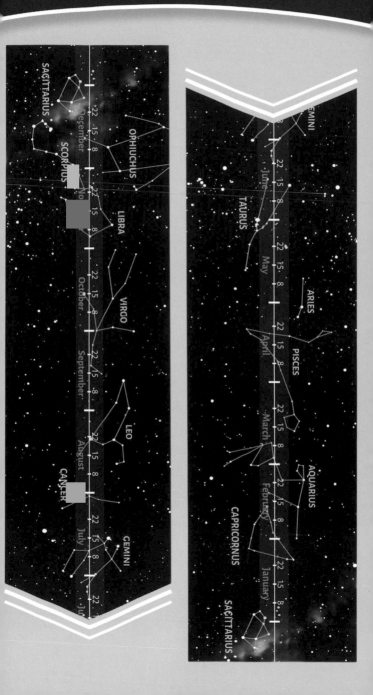

The Moon

The Moon is Earth's only natural satellite and the most prominent object in the night sky. The moon's most striking feature is its monthly sequence of phases: the moon waxes (grows) through crescent, first quarter, and gibbous (plump) phases before reaching full, then wanes (shrinks) through gibbous, third quarter, and crescent phases before disappearing again.

Full Moon

Waning Gibbous

Waxing Gibbous

Third Quarter

First Quarter

Waning Crescent

Waxing Crescent

Phases of the Moon
(When you hold this book 16 inches from your eyes, the graphic to the left shows the apparent size of the moon in the sky.)

The cycle of phases lasts 29½ days. Each quarter takes about a week. The waxing moon is visible in the evening sky, the full moon is up all night, and the waning moon is visible in the morning. Most newspapers, many calendars, and lots of apps give dates for the moon's major phases.

The surface of the moon is a mosaic of light and dark patches. Light regions are pebbled with craters from ancient impacts. Darker regions are the result of lava flows on the young moon. Large craters on the moon are named for prominent figures in the history of Western science. Dark regions were poetically named "maria," Latin for "seas," by early observers who thought they might be oceans.

Many maria and a few large craters are visible with the naked eye. Binoculars reveal dozens of craters, while a small telescope shows hundreds. The best place to observe the moon is along the "terminator"—the edge of illumination on the moon's disk. Here the sun's shallow angle casts long shadows, making surface features easier to see. During the full moon, sunlight shining straight onto the moon's visible face gives us a flat, shadowless view of the terrain.

The planets (Greek for "wanderer") move across the sky. They do not appear on the sky maps, but are always near the line of the ecliptic. Use the tips below to help identify the five bright "naked-eye" planets.

Mercury

The innermost planet, Mercury is close to the sun and never visible after full dark at northern latitudes. Although sometimes bright, it is usually obscured by twilight. Six or seven times each year, Mercury reaches its largest angle from the sun and may be visible for several days near where the sun has set or is about to rise.

Venus

Venus shines brighter than anything else in the sky, besides the sun and the moon. Over the course of a year and a half, Venus cycles between the morning and evening skies. Shining brilliant white, it is often the first "star" visible in the evening or the last visible after dawn.

Mars

Mars shines a distinctive red-orange. With an orbit just a little larger than Earth's, its distance from us, and how bright it looks, varies more than any other planet. Usually Mars is rather far away and appears dimmer than the brightest stars near the ecliptic. Once every other year, Mars passes close to Earth and shines brighter than any star.

Jupiter

Jupiter is striking in the sky, shining brighter than any star. Jupiter takes nearly 12 years to orbit the sun, spending about a year in each constellation along the ecliptic. It is a treat to view in binoculars, which may reveal the planet's four largest moons.

Saturn

Saturn has a yellowish tint that is unique among objects near the ecliptic. A very steady hand with good binoculars will show that Saturn isn't a uniform disk, although you need a telescope to resolve its famous rings.

DEEP-SKY OBJECTS

Milky Way: Our home galaxy is made up of some 200 billion stars. The collective light of these billions and billions of stars blends together in a band of light that stretches across our sky. Striking under clear, dark skies, the Milky Way is shown on each of the star maps.

Andromeda Galaxy (M31): The most distant object visible to the naked eye—2.5 million light years from Earth. To the naked eye, it appears as a faint, fuzzy oval. A spiral galaxy similar to our own Milky Way, it contains an estimated 200 billion stars and is 120,000 light years across.

Orion Nebula (M42): A diffuse nebula visible as a soft patch of light near the tip of the "sword" that hangs from Orion's Belt. It is one of the best studied star-forming regions in the sky. Binoculars show some shape and structure while a small telescope reveals beautiful texture and many young stars.

Praesepe (M44): Visible on clear nights as a faint, fuzzy circle about three times the diameter of the full moon. Binoculars reveal a dense bunching of stars that seem to swarm like bees, giving it the nickname "the Beehive Cluster."

Pleiades (M45): The most famous star cluster in the sky. Easily seen with the naked eye and spectacular in binoculars. A person with good eyesight can see six or more stars clustered in a fuzzy patch about four times the diameter of the full moon. Binoculars show that the cluster's six brightest stars form a tiny dipper.

METEOR SHOWERS

Poetically called "shooting stars," meteors are small pieces of space debris burning up in Earth's atmosphere. Although meteors may be seen at any time, there are a few times each year when the Earth passes through a cloud of debris and dozens of meteors may be visible every hour. These events are called meteor showers.

Name and Where to Look	Dates	Peak	Per Hour	Strength
Quadrantids	Jan 1–Jan 10	Jan 4	120	Strong
Eta Aquarids	Apr 19–May 26	May 7	55	Moderate
Perseids	Jul 13–Aug 26	Aug 12	100	Strong
Orionids	Oct 4–Nov 14	Oct 22	25	Moderate
Geminids	Dec 4–Dec 16	Dec 14	120	Strong

Helpful Resources

Whether you want to spot a satellite or know where to go to avoid light pollution, check out the following helpful websites and organizations:

Free planet location charts from author Jonathan Poppele
www.jonathanpoppele.com

Stellarium Planetarium Software
www.stellarium.org

Sky Maps—Free printable monthly star charts
www.skymaps.com

Heavens Above—Satellite predictions
www.heavens-above.com

Clear Dark Sky —Light pollution map
http://cleardarksky.com

The National Aeronautics and Space Administration
www.nasa.gov

Globe at Night—Citizen-science project
www.globeatnight.org

Staracle—Free site to name a star for someone
www.staracle.com

Your Night Sky Companion

Simple and convenient—choose your season, and discover the many wonders of stargazing

- Topics organized by season for quick and easy identification

- Six full-sky charts for locating constellations any time of year

- Professional photographs and illustrations

- Easy-to-follow observation tips and instructions

- Information about spotting planets, meteor showers, and more

- Pocket-sized format—easier than laminated foldouts

Improve your stargazing skills with the *Night Sky* field guide and playing cards

ISBN 978-1-59193-581-0   **$9.95**

PUBLICATIONS
Adventure
an imprint of AdventureKEEN
SCIENCE/ASTRONOMY